Marcelo Cardinale Branco

Vehicle Inspection

AF666423

Marcelo Cardinale Branco

Vehicle Inspection

As an Instrument for Controlling Air Pollution in Large Cities

ScienciaScripts

Imprint

Cover image: www.ingimage.com

This book is a translation from the original published under ISBN 978-3-330-76999-1.

Publisher:
Sciencia Scripts
is a trademark of
Dodo Books Indian Ocean Ltd. and OmniScriptum S.R.L publishing group

120 High Road, East Finchley, London, N2 9ED, United Kingdom
Str. Armeneasca 28/1, office 1, Chisinau MD-2012, Republic of Moldova, Europe
Managing Directors: Ieva Konstantinova, Victoria Ursu
info@omniscriptum.com

Printed at: see last page
ISBN: 978-620-8-62713-3

SUMMARY

"The burning of fuels - mainly fossil fuels - still represents a preponderant share of the source of the energy used by human beings in their activities.

However, this practice has serious consequences for the environment, including air pollution, the greenhouse effect and acid rain. Most countries are making efforts to replace climate-damaging fuels with less harmful energy alternatives."

Samuel Murgel Branco (in Energy and Environment)

SUMMARY

The reliable and safe movement of people is fundamental to the development of any economy, playing a key role in educational, commercial and cultural issues. However, it is the transport sector that is largely responsible for the emission of polluting gases into the atmosphere of large urban centers today, not just in Brazil, but throughout the world. The large-scale use of energy inherent in this sector is the major source of emissions of polluting gases and heat. Over time, most (if not all) of São Paulo's industrial sectors have significantly reduced the volume of pollutant emissions in their production processes. However, achieving a large-scale reduction in pollutant emissions from vehicles is a much more complex task, as there are almost 6 million small "industries" to be controlled in the city of São Paulo alone. In this context, it is therefore necessary to discipline the automobile industry so that vehicle models with higher performance engines and systems and successively lower emissions come out of the factory again and again, and to oblige the correct maintenance of vehicles in circulation, guaranteeing the quality of the fleet. Balanced modal distribution in motorized transport (how people are transported, in private or collective vehicles) is also a very important tool in this mission to reduce the volume of atmospheric pollutants from vehicles, since the volume of emissions per passenger transported in a bus, for example, is significantly lower than that observed in a car or motorcycle. And the greater the capacity of the public transport equipment, the better this ratio will be, for a given technological standard. The aim of this paper is to analyze pollutant emissions in the transport sector in the city of Sao Paulo, a pioneer in drawing up and complying with its PCPV - Vehicle Emissions Control Plan, showing that an efficient system for reducing pollutant emissions from vehicles must include three fundamental aspects: the search for a modal balance; control of the environmental quality of the vehicles manufactured; and control of pollutant emissions from the fleet in circulation.

1 INTRODUCTION

The transportation of people has been a topic of the utmost relevance in important discussion forums on city management around the world.

The "habit of travel", a term widely used in technical documents dealing with issues related to transport systems in the first half of the 20th century, was defined as the number of trips made by a municipality in a given period of time.

This way of measuring the movement of people in a city was later replaced by the concept covered by the term "Mobility" which, despite having a similar meaning in the dictionary, as "Displacement of individuals, groups in the urban space" (Michaelis, 2010), its scope has been broadened in current works dealing with the subject, taking a more comprehensive view than that first term, meaning not only the number of journeys actually made by citizens in a given city or region, but also the possibility of making those journeys.

Thus, if measuring the "habit of traveling" was considered sufficient in times past, today's public administration sectors, whether federal, state or municipal, directly linked to the transportation of people, must be concerned with providing transportation systems that allow the number of trips desired by the citizen to be made, otherwise they will be measuring the number of trips made, without identifying that the desire or need to travel would be much greater than what was possible.

In order to meet this desire to travel, providing mobility in large urban centers is a challenge for public administrators, mainly due to the large amount of resources needed to implement and maintain efficient transport systems, especially medium and high capacity ones, which are not always available, particularly in cities in developing countries.

While this need for mobility, the movement of people and products, consumes a large part of the budgets of Brazil's big cities, existing projects have proved insufficient to meet social demand.

Over the decades, Brazil has adopted a road-oriented culture, directly or indirectly encouraging the use of the car, individual transportation, to the detriment of collective systems.

This story comes at a cost: in addition to the huge traffic jams seen today not only in Brazil's big cities, but which is a phenomenon that has already been widely identified in medium-sized and small towns, the volume of energy used in the form of fuels is huge, generating, at the other end, enormous quantities of pollutants, released into the atmosphere.

In addition to the huge shortage of resources allocated to building, operating and maintaining transport systems, the lack of such equipment leads us to pay a huge social cost, as the amount of time and fuel spent and the wear and tear on existing transport equipment are high due to the slowness and overcrowding of the systems, This is further aggravated by the huge traffic jams in large cities, which are also largely the result of the lack of public transport that meets the population's needs in terms of quality and price. This has created a perverse vicious circle, where the lack of adequate public transport leads people to look for other means of transportation, crowding the streets with cars and motorcycles which, by creating congestion, contribute to reducing the speed of buses and public transport, which becomes even less attractive.

This intense use of the road system is still largely responsible for the increase in the volume of emissions of polluting particles and gases from cars, buses, trucks and motorcycles, and for the creation of traffic jams. Compared to free-flowing stretches of road and the use of constant gears, a vehicle idling, driving slowly in low gear, constantly changing gears or stopped in a traffic jam consumes much more fuel and emits a greater volume of pollutants per kilometer traveled.

In other words, if on the one hand we don't have the volume of resources needed to set up transport systems that meet the needs of the cities, on the other we see a much larger volume of resources wasted on congestion and the reduction in worker productivity (the average loss of worker productivity in the big cities of Latin America, including Sao Paulo, was estimated at 5% due to their exposure to congestion - Citigroup, 2013), including Sao Paulo, was estimated at 5% due to their exposure to congestion - Citigroup, 2013) and the damage caused to society by the sector's shortcomings and the excessive use of individualized transport, cars and motorcycles, is increasingly damaging to countries.

In parallel with this great difficulty in setting up public transport systems, there is an understanding on the part of public administrators that the construction of roads for the free circulation of individual transport vehicles is a great promoter of popular support.

This understanding has led countless administrations to invest the always insufficient public resources in the construction of roads and works of art aimed primarily, if not exclusively, at the circulation of automobiles and individual transportation, not to mention the lack of urbanistic commitment in the development of the projects for these major works, which in no way contribute to the sustainable development of urban centers.

Associated with this mistaken model, which believes that the solution to mobility in large urban centers lies in the expansion of individualized transport, we have seen policies that strongly encourage the automobile industry, including tax exemptions and reductions, the abundant resources made available to finance vehicles and the huge and constant subsidies given to fossil fuels, especially gasoline, creating an environment conducive to the huge increase in the number of vehicles in all Brazilian cities in the last decade.

This whole reality has led Brazilian cities to a huge modal imbalance, a large concentration of journeys made by motorized individual transport equipment, cars and motorcycles, which in turn leads to an excessive volume of pollutant emissions and energy expenditure.

As a result, the volume of cars and motorcycles is now extremely significant in our transport matrix, and it is important to analyze how much pollutants these vehicles generate and what mechanisms are in place to control these emissions.

And if the excessive use of cars in cities leads to such undesirable consequences for society, it is also necessary to analyze the individual environmental quality of the vehicle that circulates on city roads.

In this respect, on the one hand, it is essential for manufactured vehicles to be controlled in terms of pollutant emissions and, in addition, for the quality of vehicles in circulation to be controlled. In other words, manufactured vehicles must follow increasingly restrictive pollutant emission parameters, polluting less with each new launch, and must remain so throughout their useful life, not allowing the ageing of the vehicle to mean excessive deterioration in its environmental quality, with a consequent excessive increase in emissions.

All of these issues lead us to one of the main global concerns of today: the environmental impacts caused by transportation systems, mainly generated in large cities, but which are damaging to all of humanity due to their global effects.

The World Health Organization (International Agency for Research on Cancer - WHO - Report October 2013) has classified air pollution generated by the burning of fossil fuels as one of the main causes of lung cancer, showing that the concentration of these pollutants is more damaging to the public health system than the use of cigarettes. According to the report, more than 220,000 people died in 2010 from lung cancer caused by particulate matter, hydrocarbons and aldehydes in the atmosphere from burning fossil fuels.

Air pollution alone is responsible for around 3% of the world's deaths from cardiopulmonary diseases, 5% of lung cancer cases and 3% of deaths from respiratory infections in children under five years of age. This corresponds to approximately 800,000 deaths per year (SALDIVA et al, 2010).

In recent decades, this concern about the pollution generated by the burning of fossil fuels, including the emission of greenhouse gases, has been demonstrated at international events, such as the First World Climate Conference in Geneva, Switzerland, in 1979, which launched the United Nations Environment Program; the Rio 1992 conference, promoted by the UN and which proposed Agenda 21, and the Kyoto World Conference in 1997, which resulted in a global protocol proposing the reduction of greenhouse gas emissions for the following decades.

The concern is justified because not only are the polluting gases highly damaging to people's health and quality of life, but the massive use of energy has the effect of generating heat, contributing to a rise in the ambient temperature of cities and the generation of heat islands. In other words, regardless of where they are consumed, the use of these fossil fuels and their effects becomes a global issue and not just a local one.

But while on the one hand the effects of fossil fuel consumption are extremely worry-worthy, on the other hand we can see that the country has made great strides in controlling pollutant emissions, when it comes to the environmental quality of cars, with the creation of the Program for the Control of Air Pollution by Motor Vehicles - PROCONVE, in 1986.

This program disciplined the national automobile industry from the point of view of pollutant emissions generated by new motor vehicles.

Since then, the average levels of pollutant emissions from motor vehicles have been gradually falling, both for vehicles using ethanol and/or gasoline engines and for diesel vehicles, in some cases with reductions of more than 95% in emissions over time (ENVIRONMENTALITY, 2012).

Naturally, this reduction in emissions generated mainly by the change in vehicle technology to limit the maximum levels of pollutants authorized by PROCONVE for each make, type and year of vehicle leaving the factories is reflected, through fleet renewal, in the levels of pollutants emitted by vehicles in circulation.

The benefits generated by PROCONVE can be exemplified (CONTROLAR IM/SP annual monitoring reports - 2008 to 2012): in the measurements made by the Inspection and Maintenance Program for Vehicles in Use in the Municipality of São Paulo (I/M-SP), vehicles approved for inspection (in a first test or re-inspection) showed between 0.05 and 3% CO emissions at idle speed, depending on the year and PROCONVE phase in which the vehicle was manufactured. In other words, vehicles manufactured before 1988 (known as pre-PROCONVE) have a CO emission level 60 times higher in the idling test than those manufactured from 2009 onwards (PROCONVE phase L5). The same goes for diesel vehicles, which over time have seen a sharp reduction in the volume of particulate matter expelled from the exhaust. While for vehicles manufactured before 1988 (pre-PROCONVE) that passed the municipal test, the average emission of particulate matter at free acceleration was 180 mg/Nm3, for more modern vehicles (PROCONVE phase P5, manufactured from 2006 onwards) this average was 95 mg/Nm3 (SVMA, parameters for mandatory vehicle inspections, 2010).

As a result of this gradual reduction in pollutant emission limits and annual control, it can be seen that the level of pollutants in the atmosphere has been reduced, especially in cities that have implemented effective actions to control vehicle emissions, as provided for in PROCONVE (take the city of São Paulo, for example, where the number of days with inadequate air quality as measured by CETESB has been drastically reduced over the years, This is due to the progress made by PROCONVE, in which the limits for the emission of pollutants by vehicles sold by automakers were restricted, and is accentuated in the period that coincides, not by chance, with the implementation of the municipal PCPV (Vehicle Pollution Control Program), This demonstrates that the environmental gain only becomes permanent if this vehicle, which is equipped with increasingly modern technology, is properly maintained.

It is the government's duty to ensure air quality and to certify that each vehicle in circulation maintains the appropriate pollutant emission parameters. This is done through the vehicle inspection mechanism, which should ideally be carried out at the same frequency as the mechanical overhauls recommended by vehicle manufacturers.

Based on this scenario, the proposed study aims to analyze the issue of pollutant emissions produced by motor vehicles in the city of São Paulo.

The contribution of the study will be to demonstrate the need for an integrated emissions control policy, which addresses the environmental quality of the vehicle manufactured, the maintenance of these emissions factors obtained by the automobile industry through mechanisms for periodic inspection of the fleet in circulation and, finally, balancing the transport matrix, concentrating efforts on expanding quality public transport systems, which can reduce the use of individual transport and, consequently, the volume of emissions per passenger transported in the average municipal transport matrix.

The main objective of the work will be to analyze the following central questions on this subject:

- Average individual pollutant emissions associated with vehicles produced in recent decades in Brazil;

- Compare the amounts of energy used by each mode of transport in relation to one passenger transported;

- Promote an analysis of the measures that have proved effective in reducing these emissions of vehicular pollutants;

- Benefits of controlling the environmental quality of vehicles in the automotive industry - PROCONVE;

- Benefits of quality control of preventive maintenance on vehicles in circulation - I/M programs.

- To analyze whether the association of PROCONVE with São Paulo's vehicle inspection program has brought about an effective reduction in the concentration of atmospheric pollutants in the city.

2. CONSIDERATIONS ON MODAL DISTRIBUTION IN THE CITY OF SÀO PAULO

Individual transport has some advantages over public transport that are highly valued by its supporters.

The ease of getting around, with access to any region of the city or even outside its boundaries, is pointed out by the population in general (CET, 2012) as the greatest benefit offered by this mode. The speed and convenience of origin and destination, from within the garage of the home or workplace, with an equally convenient destination for the user, is also pointed out in preference surveys carried out by CET/SP in 2012, used to guide mobility and safety projects on the municipal road. The same survey indicated, with a lesser degree of importance, greater security against theft and aggression as a determining factor in the preference for individual transportation.

Data from the São Paulo Municipal Transportation Department (SMT) from 2011 indicates that in the city of São Paulo, the overall average number of people transported per individual vehicle unit (considering only cars and vans) is 1.6 passengers per vehicle. In other words, transporting 190 passengers, the maximum volume of people transported in a bi-articulated bus, would require, on average, just under 119 cars. The same number of passengers could be transported by two standard buses, with much lower emissions and energy costs per passenger, since an articulated bus travels an average of 1.5 km per liter of fuel, while the average distance traveled by a passenger car is around 9.0 km per liter of fuel used.

The city of Sao Paulo once had its citizens' transportation based on public systems.

In the mid-1970s, in association with the State Secretariat for Metropolitan Affairs (later replaced by the current Secretariat for Metropolitan Transport in matters related to mobility in metropolitan regions) and the São Paulo City Council, a plan called SISTRAN was drawn up, which strongly favored public transport, pointing out the need to set up a metro-rail network suited to transport demands, expanding the metro network, creating a company responsible for operating metropolitan trains and setting up a trolleybus network to operate in exclusive corridors, with a forecast of 280 km of network and 1280 cars, including articulated and simple vehicles. (BRANCO, 2008)

The choice to invest heavily in electric vehicles at the time was due to the high cost of fuel and the already visible need to reduce pollutant emissions and noise generated by combustion engines.

Over the decades, however, public transport has given way to individual vehicles, to private transport. Literally giving way, with the construction of streets and avenues where public transport is now competing for space with cars and motorcycles.

In 1997, the number of people transported by public systems (buses, trains and the metro system) and individual transportation (cars and motorcycles) reached a point of equality. At present, the public system is responsible for just over half of this total volume of passenger transport, as can be seen in figures 1 and 2, based on data taken from the latest Origin and Destination survey (OD 2007), drawn up by METRO/SP:

Figure 1- Average daily number of trips by mode (in millions)

	Year	1967	1977	1987	1997	2007
Mode						
Collective		4,90	9,8	10,5	10,5	13,8
Individual		2,3	6,2	8,3	10,1	11,2
Total motorized		7,2	16,0	18,8	20,7	25,0

Source: Metro: Annual Report 2007 - Origin and destination survey - STM/GESP

Figure 2- Share of total motorized trips, by mode

	Year	1967	1977	1987	1997	2007
Mode						
Collective		68,1%	61,0%	55,8%	50,8%	55,0%
Individual		31,9%	39,0%	44,2%	49,2%	45,0%

Source: Metro: Annual Report 2007 - Origin and destination survey STM/GESP

This percentage distribution of passengers transported in public and private systems observed in recent years remained at the same levels in the partial checks carried out between 2007 and 2012, with the number of passengers using public transport exceeding the number of passengers using individual transport. However, this new level is not the result of a desirable reduction in the number of vehicles on the streets or in the number of daily journeys made by these vehicles. On the contrary, the number of vehicles registered in the city of São Paulo has grown significantly. The big increase in the number of journeys on public transport was due to the implementation of the timed ticket on the city's bus system (bilhete único), which added more than 3 million journeys a day to the public system between

1997 and 2007, while the number of journeys by car increased by just under 1 million journeys a day over the same period of time.

Figure 1 shows that in recent years there has been a sharp increase in the total number of motorized journeys in the city, mainly as a result of the implementation of the single ticket, as mentioned above, and the integration of fares between the various public systems (trains, subways and buses).

This time-based and integrated fare format has led to an increase in the average number of journeys per passenger using the system, insofar as it allows multiple journeys of up to three hours on the bus system and the payment of only part of the fare when the user uses metro rail transport in addition to bus transport.

The data shows that in the 1960s the city of Sao Paulo had around 70% of all journeys made by public transport systems, with only 30% of motorized journeys made by individual means.

Of course, all the systems have changed a lot since then, with the growth of the city and the mobility of people, as well as the circulation of goods and the volume of services, much greater access to the purchase of vehicles by the population, in addition to the implementation of the metro system in the city.

The current modal distribution is the result of a sum of public policies, from the various spheres of government, which has allowed cars to occupy public spaces, to the visible detriment of surface public transport.

The redistribution of the use of this public space must be planned in such a way as to attract users to the public transport system, with good designs and proper operation, offering quality transport at a cost that is compatible with the ability of the poorest sections of the population to pay, with comfort and reliability, otherwise we will end up with huge investments in transport systems, often hampered by congestion caused by cars (for example, bus lines that compete with cars for the use of the same lanes on the road) totally or partially idle, if they are not able to compete with the attractiveness of individual transportation which, even though it demands greater volumes of energy, costs more and generates greater emissions of pollutants, still seems unsurpassed in the preference of Brazilian citizens.

With the widespread use of the streets by individual cars and motorcycles over time, it would be necessary to implement high-capacity elevated or underground transport to replace the old streetcars and, later, bus systems, so that the population would have an alternative to the individual option.

Naturally, this implementation has not been in line with the needs of the population, which, combined with policies to facilitate the purchase of vehicles (cars and motorcycles), leads citizens to the mistaken view that they will only achieve their desired mobility if and when they manage to acquire a car or motorcycle, thus creating a vicious circle that is extremely damaging and contrary to quality of life and better use of public space.

This phenomenon of increasing use of individual means of transport is not only happening in Brazilian cities.

This is, in fact, another issue that concerns city administrations in various countries, despite the fact that public transport accounts for a more significant share of the transport matrix in European cities (in Stockholm, for example, 70% of motorized journeys are made on public transport systems, while individual transport accounts for 30% of the total), the trend towards greater use of the car is a constant, making it necessary to set up a wide-ranging transport network that allows people to access public transport anywhere in the city with short journeys on foot, in order to counter the almost free origin and destination convenience offered by individual transport.

The document entitled "White Paper: Transport Policy for Horizon 2010: Time to Choose", a document published in 2001 by the European Union, warned of the strong trend towards the use of individualized transport in European countries and the concern of the group of countries about this trend, both in urban movements and in road travel, showing that the cost of traffic jams and accidents in the cities of the countries that make up the European Union was running at around 1% of the Union's Gross Domestic Product and was also responsible for 28% of the volume of CO emitted by the countries as a whole.

These emissions increase in line with the use of fuels, when comparing similar technologies. From the point of view of the type of emissions, it is difficult to compare public transport in Brazil (100% diesel buses) with other modes (individual transport essentially ethanol, gasoline), because each fuel is associated with specific pollutants (diesel - particulate matter and nitrogen oxides; gasoline - CO; ethanol - hydrocarbons), even though some pollutants are associated with the burning of all fuels, in different proportions.

However, the comparison of polluting potential is made possible by the relationship between energy used and the volume of passengers transported (Environmentality - 2013), since the emission of pollutants is a direct consequence of the incomplete burning of fuel components and is therefore directly related to the quantity used, comparing similar technologies.

In the transportation system as a whole, the use of fossil fuels has increased over the years (source: ANP 2012 annual report), showing that the demand for energy grows with the increase in the fleet, as there are no consistent and mandatory targets for increasing energy efficiency in the sector.

The key to increasing energy efficiency in transport lies in modal shift, using less energy and therefore less fuel per passenger transported and, as with the issue of pollutant emissions, in technological change and increasing the efficiency of engines or traction systems.

Figure 3 shows that the ratio of consumption proportional to one passenger transported is frankly favorable to public transport systems, reinforcing the understanding that modal distribution can be a powerful instrument in reducing emissions of polluting gases in the sector, in addition to the fact that burning less fossil fuels is in the country's economic and strategic interest.

The use of electrically-powered vehicles such as trolleybuses also has the advantage of near-zero emissions (non-zero due to the emission of pollutants not directly related to powering the engine, such as particles from brake linings, tires, etc.) when operating from the vehicle's power supply.

Figure 3- Energy consumption by mode of transport

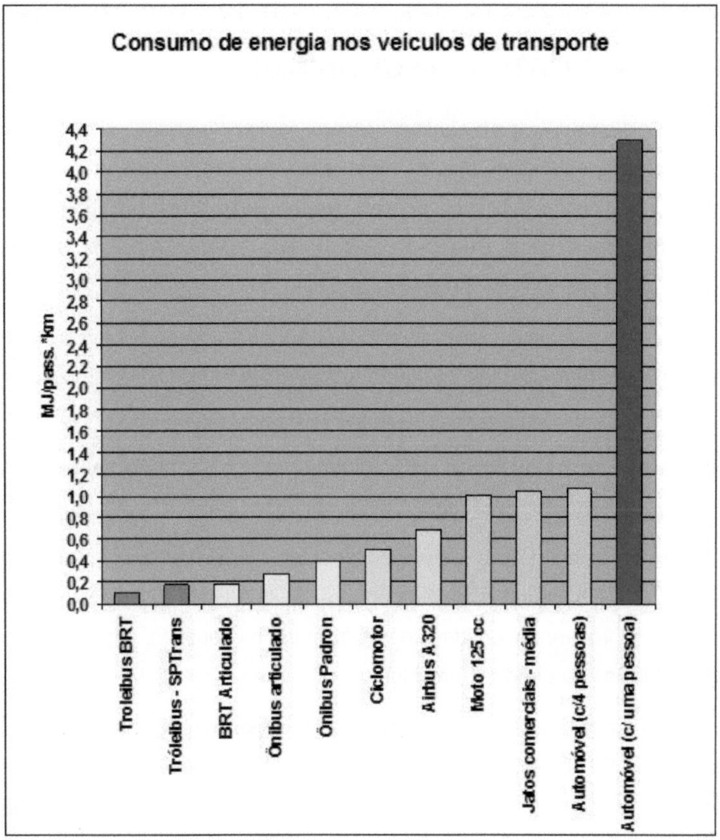

Source: Environmentality, 2010

3. POLLUTANT EMISSIONS IN THE TRANSPORT SECTOR AND REDUCTION AND MITIGATION MECHANISMS

The transportation sector is currently the biggest contributor to air pollution in the city of São Paulo and its metropolitan region. In fact, in most analyses of the concentration of atmospheric pollutants, it is not possible to separate the origin of these polluting gases between the cities of the Metropolitan Region of São Paulo - RMSP, which is why the analysis of air quality is done together for the entire RMSP.

However, several studies on air quality in the city of São Paulo have been carried out over the last 10 years with the aim of measuring the damage caused by atmospheric pollution to people's health.

It is estimated that in two years more than 20,000 children and 8,000 elderly people have been hospitalized for respiratory problems related to air pollution (CONTROLAR, 2012).

The use of diesel fuel in buses and trucks is responsible for the largest share of particulate matter and nitrogen oxides (NOx) found in the atmosphere of cities, while 75% of the volume of carbon monoxide and hydrocarbons emitted comes from the combustion of gasoline and ethanol used in light vehicles (cars and motorcycles).

If the transport sector is known to be largely responsible for these volumes of pollutants, it is essential to implement public policies aimed at reducing these emissions, either by promoting technological improvements in the manufacture of vehicles of better environmental quality by industry, or by controlling the emissions of vehicles in circulation, with requirements that bring vehicles back into line with their original certified specifications, or by working on modal distribution.

Analyzing air quality in the city of São Paulo and its metropolitan region, we can clearly see that there has been a negative trend in the number of days of inadequate air quality coinciding with two distinct periods: the implementation and evolution of PROCONVE and the creation and implementation of the Vehicle Pollution Control Plan - PCPV in the city, including vehicle inspections. In other words, these two major environmental programs, one of them at national level, but with measurable results at local level, and the other at municipal level, were responsible for reducing the number of days with air considered by the Environmental Sanitation Technology Company (CETESB) to be of inadequate standard.

For a more precise analysis, it is important to note that air quality indices are strongly related to factors of atmospheric dispersion of pollutants, i.e. although they are directly related to the total volume of pollutants released into the air in a given period of time, the concentration and consequent quality of the air depend on wind, rain and other atmospheric factors capable of dispersing the pollutants.

Figure 4 shows the evolution of the air quality indices for CO, indicating the percentage of days when the air quality was not good and the number of days unfavorable to the dispersion of atmospheric pollutants reported by CETESB in its annual reports on air quality in the São Paulo Metropolitan Region, showing that the variation in the concentration of pollutants in different dispersion conditions is quite significant, but also that the concentration of these pollutants is, despite the increase in the number of days with unfavorable conditions for the dispersion of pollutants, always at levels significantly lower than those perceived in years prior to the emission control programs mentioned. In other words, despite the fact that atmospheric dispersion significantly modifies air quality, the reduction in the level of pollutants in the air in the period after the implementation of the PCPV in the city of São Paulo is clear, compared to the period before the program.

Carbon monoxide was taken as an indicator because it is the pollutant that best indicates the influence of vehicles on air quality for two main reasons: it is emitted almost exclusively by vehicle traffic and is therefore highly dependent on this emission source; secondly, because its sample is an 8-hour moving average (and not 24 hours like the other indicators) and demonstrates its relationship with peak traffic times in the city.

In the graphical demonstration, "not good" was considered to be air quality that was not in the "good" or "excellent" range according to CETESB's assessment.

Figure 4. air quality indices in the RMSP - carbon monoxide

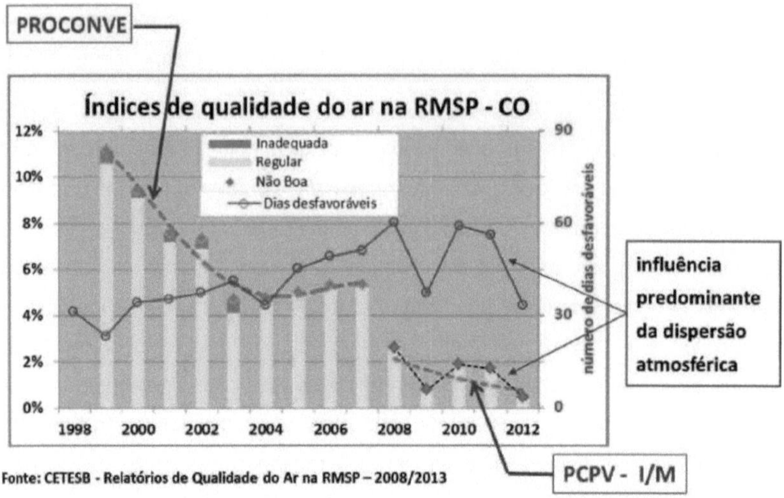

Fonte: CETESB - Relatórios de Qualidade do Ar na RMSP – 2008/2013

Two main pieces of information are shown in figure 4: the yellow and orange bars show, as a percentage, what portion of the year the CO concentration levels reached regular (yellow) or inadequate (orange) levels. The sum of days in these conditions is represented by the green dots ("not good" condition). This percentage is scored on the left axis of the graph.

The other piece of information is the number of days when the atmospheric dispersion conditions were unfavorable. This is represented by the red line.

The purpose of putting these two pieces of information in the same figure is to show that the concentration of atmospheric pollutants follows the dispersion conditions in a fairly constant way. However, it is also clear that, while the dispersion of pollutants takes place to a greater or lesser extent depending on atmospheric conditions, the concentration of these pollutants depends heavily on public policies, and the reduction of these pollutants in the atmosphere is clear from PROCONVE (until 2003) and the implementation of PCPV, including vehicle inspections in the city of São Paulo.

The concentration of atmospheric pollutants is, of course, directly related to the number of vehicles of a certain technological standard in circulation. The vehicle fleet in the city of Sao Paulo has grown steadily over the years, as shown in figure 5, as a result of the increase in consumption in the country and policies to encourage the purchase of cars and motorcycles, especially in recent years.

In this way, it can be attested that the effective improvement in air quality was due to the increase in the environmental quality of the fleet in circulation, with a lower average emission of pollutants per vehicle, since the number of emitters (vehicles) grew by more than 70% in the period between 2001 and 2013.

Figure 5: Percentage increase in the number of vehicles registered in the RMSP

YEAR	FLEET GROWTH IN RELATION TO TO THE PREVIOUS YEAR
2002	4,6%
2003	4,0%
2004	3,8%
2005	5,3%
2006	4,8%
2007	7,5%
2008	7,6%
2009	5,8%
2010	4,1%
2011	3,6%

Source: DENATRAN, 2013

The reduction in the number of days with inadequate air quality in the city of Sao Paulo and the metropolitan region, therefore, was due solely to environmental control of the vehicles produced and the fleet in circulation.

PROCONVE's requirements obliged the industry to develop or import and implement devices such as direct fuel injection, fuel gas evaporation control and catalytic converters, among others, increasing the environmental quality of the vehicles produced.

On the other hand, the compulsory vehicle inspection of the public and private fleets, requiring proper maintenance to bring the pollutant emission parameters of these vehicles in circulation closer to those predicted by the automakers, considering normal wear and proper maintenance, has not allowed these vehicles to deteriorate excessively. In other words, the greater number of vehicles in circulation was largely offset by the better environmental quality of this fleet compared to previous years.

The government programs responsible for this reduction were PROCONVE and I/M-SP, described in the following chapters.

4. HISTORY AND BENEFITS OF THE PROGRAM FOR THE CONTROL OF AIR POLLUTION BY MOTOR VEHICLES - PROCONVE

PROCONVE was created in 1986 to regulate the production of motor vehicles in terms of the emission of polluting gases and noise. In 2002, it was complemented by the Program for the Control of Air Pollution by Motorcycles and Similar Vehicles (PROMOT), which defines pollutant and noise emission standards for motorcycles, expanding control over the manufacture of motor vehicles.

Since its creation, the automobile industry has been producing vehicles that are obliged to follow maximum emission standards previously established and gradually reduced over the years and stages of the Program.

According to IBAMA's website (www.ibama.gov.br, on 10/02/2014), "*the increase in individual motorization, resulting from the chronic deficiency of mass transport systems, has intensified traffic in large urban centers. In addition to causing constant congestion, with the consequent environmental degradation due to air and noise pollution caused by motor vehicles, the growth in the number of vehicles increases socio-economic costs and causes serious damage to human health, which must be controlled through the adoption of effective measures to control vehicular pollution, directly or indirectly. With the aim of reducing and controlling atmospheric pollution from mobile sources (motor vehicles), the National Environment Council (CON AMA) created the Programs for the Control of Air Pollution from Motor Vehicles: PROCONVE (cars) and PROMOT (motorcycles), setting deadlines, maximum emission limits and establishing technological requirements for national and imported motor vehicles*". These limits and requirements are periodically updated.

Over the course of its more than 25 years of existence, PROCONVE has led to a drastic reduction in the volume of pollutant and noise emissions from the fleet of vehicles in circulation, to the extent that it has controlled the production and import of vehicles with regard to environmental aspects. Over time, older vehicles are gradually replaced by more modern ones, with much stricter emission standards, leading to a reduction in air pollution in large cities, where there is a higher concentration of these pollutants and more people are exposed to them.

PROCONVE's objectives are not restricted to reducing the levels of pollutants emitted by motor vehicles, but also to promoting technological advances in the automotive industry and in the mechanisms (equipment and methodologies) for controlling emissions, as well as the development of fuels.

These issues are closely linked. In order to implement technological advances in engines, it is necessary for fuels to evolve. As an example, engines with the technology to reduce emissions to the Euro 5 standard, which is the fifth stage of the European vehicle emissions control program, must necessarily use diesel fuel with less than 50 ppms (parts per million) of sulphur, at the risk of malfunctioning of the gas aftertreatment system and mechanical damage to the engine. In other words, diesel vehicles with engines imported from Europe from 2010 onwards can only use diesel with less than 50 ppms of sulphur, which must therefore be available at petrol stations.

The creation and implementation of this program has led the automobile industry to incorporate technological advances into its vehicles that allow them to comply with the legal limits for gas and noise emissions. Therefore, the environmental control program has induced this technological advance up to this point and the new phases of PROCONVE to be launched will certainly force the industry to make further advances, since the Program is constantly evolving.

Looking at the technological issue of vehicles as alternatives to the current traction models used, we have already seen widespread use of hybrid fuel/electric and purely electric vehicles in other countries.

According to Toyota, more than one million hybrid fuel/electric Prius vehicles have been sold in the United States alone. Around the world, 6 million Prius vehicles have been sold in 80 countries.

In the city of Sao Paulo, these technologies have already been used in public transport vehicles, mainly with the implementation of electric vehicles, used in trolleybuses since the 1960s and completely modernized in recent years, and in the experiment that implemented the fleet of hybrid fuel/electric and pure electric taxis in the city, The electric ones are experimental, using the Nissan Leaf vehicle, not yet sold in Brazil, and the hybrid ones are already at the commercial stage, using Toyota Prius vehicles, identical to those available for sale in the dealer network.

Taking the Toyota Prius as an example, since it is a production vehicle available for sale in dealerships, the reduction in fuel consumption was quite marked in the tests carried out by the Municipal Department of Transportes de Sâo Paulo (SMT) in 2012. While a similar vehicle of the same brand, powered exclusively by petrol, had an average range of 9.4km/l, the hybrid vehicle had an average of 27km/l in urban traffic (SMT performance report/October 2012).

This reduction in consumption varies according to how the vehicle is used. Hybrid vehicles, for example, have a large advantage over a similar vehicle powered exclusively by a combustion engine when used in heavy traffic, the so-called "walk and stop", due to their mechanism for harnessing energy from braking and using this energy by the electric motor when starting off. In highway use, at higher and more constant speeds, this advantage over its technological competitors becomes less evident, as it starts to use its combustion engine more intensely.

This leads us to believe that technological progress is still the way forward for reducing vehicle emissions.

This technological evolution observed in the automobile industry worldwide indicates that, despite the strong reduction in the emission parameters allowed by PROCONVE over the last two decades in Brazil, there is a long road of technological progress to be made available in the country and the format of constantly demanding a reduction in the emission standards of new vehicles, which was the basis of PROCONVE, continues to be the appropriate way to force the incorporation of increasingly modern technologies into the vehicles sold in the country.

In other words, the continuity of the well-run program could lead the automobile industry to implement on a large scale these technologies available in other countries, otherwise they will not meet the environmental targets required in the future, reducing Brazilian emission standards even further.

This demonstrates the importance of continuing and constantly improving the automotive industry's environmental control program in the country, without which we run the risk of once again absorbing obsolete technologies in other countries.

This progress can be demonstrated by analyzing the emission limits tolerated for each technological level of vehicle, which have become more restrictive over the years, and by demonstrating the reduction in environmental impact (CO and HC) over the years in which the restrictions have been imposed.

As mentioned in the previous chapter of this paper, the vehicle fleet has grown considerably in the city of São Paulo, following the trend observed in country. Despite this, the environmental impact has been reduced, precisely because of the technologies incorporated into vehicles, as a result of the requirement to meet the targets set by PROCONVE. Figure 6 shows the emission limits tolerated by PROCONVE for each segment of the fleet (each of the technological classifications or phases corresponding to the PROCONVE stages was used as a segment).

Figure 6: PROCONVE emission limits

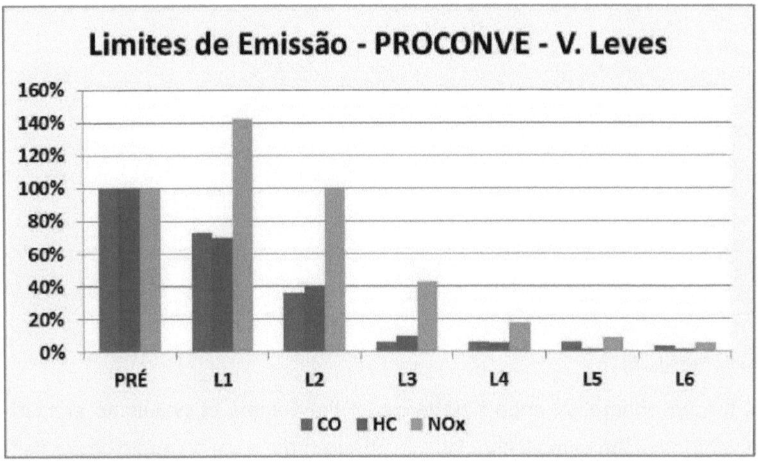

Source: (CETESB, 2012)

Interpretation:

- Figure 6 shows a comparison of the emission limits established by PROCONVE for light vehicles powered by gasoline and ethanol, based on the average emission before the implementation of the Program (100%);

- In the very first phase of the program, there was a reduction of approximately 30% in the limit for CO and HC and an increase of 40% in the limit for NOx;

- The increase in the NOx limit can be explained by the fact that although the average emissions of this pollutant were lower than the limit established at the start of the Program, many new vehicles were far exceeding this average, requiring a period of adaptation for these models. This is why, in the second phase, a limit was set for this pollutant equivalent to the average that existed before PROCONVE, while for the other pollutants - CO and HC - the reduction was constant. From then on, the reduction of parameters for the three pollutants is continuous.

Figure 7 shows the evolution of the fleet compared to the environmental impact generated:

Figure 7. Fleet and Impact - Light Vehicles

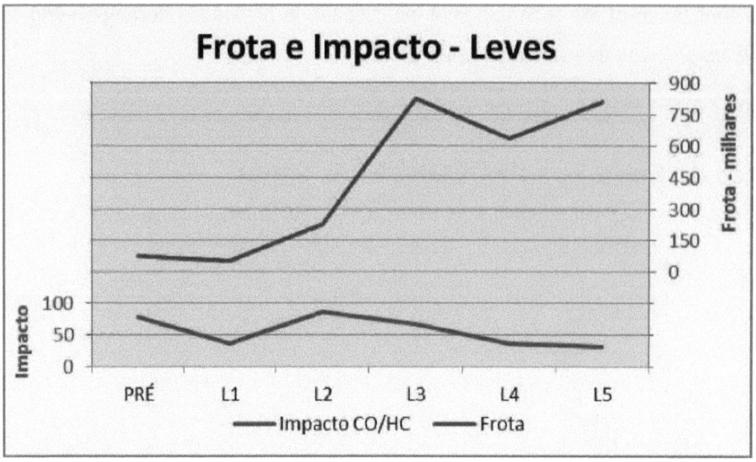

Source: (SVMA/ Annual Report, 2012)

Figure 7 shows the environmental impact (in terms of the volume of pollutants emitted) generated by all vehicles of the same technology. In other words, while an approximate fleet of 75,000 vehicles of pre-PROCONVE technology have an impact of 75 (on a scale of zero to 100 units), a fleet of more than 800,000 vehicles of L5 technology (therefore, PROCONVE phase 5) generate an impact of less than 50, on the same scale. In other words, 75,000 Pre-PROCONVE vehicles emit one and a half times the volume of pollutants (CO and HC) emitted by 800,000 L5 vehicles.

This downward trend in environmental impact in the face of a growing vehicle fleet demonstrates the success of the Program as a public policy, and it is essential that there is constant control of the automobile industry in a vehicle emission control system.

Naturally, the logical sequence for reducing pollutant emissions from vehicles must start with their production, with vehicles that have the technology to enable this environmental condition, making it technologically much more complex and socially unjustifiable to reduce emissions after the particular vehicle has been manufactured, improving its environmental condition. This mechanism, called retrofitting, is only used in large fleets, usually in public fleets or fleets of large companies (for example, some buses in the São Paulo public fleet) and in specific cases where the useful life of older vehicles is to be extended, improving their quality in environmental terms, which makes it possible to plan the renewal of the fleet in stages. But their efficiency will always be lower than that obtained in a vehicle equipped with the same devices when it was manufactured and the cost will invariably be higher. (analysis based on data from the document Ecofrota: Possibilidades de Retrofit -

SMT/2011).

5. VEHICLE INSPECTION - THE EXPERIENCE OF THE CITY OF SÂO PAULO

The annual inspection of vehicles in circulation - I/M - was one of PROCONVE's objectives in CONAMA resolution 18/86 and established by CONAMA resolution 07/93, which was later extended and replaced by CONAMA resolution 418/2009, which also included control requirements for motorcycles and similar vehicles and established procedures for measuring exhaust noise, and was supplemented and updated by CONAMA resolution 451/12.

It is important to note that the inspection and maintenance mechanism for the fleet in circulation is provided for in PROCONVE and PROMOT, i.e. it is an integral part of those federal control programs, so it is not admissible for these inspection and maintenance monitoring mechanisms to be drawn up without observing the rules established in PROCONVE and PROMOT, such as annual inspections.

The City of São Paulo implemented its Motor Vehicle Inspection and Maintenance Program (I/M-SP) in 2008, with the structure of inspection centers gradually being put in place.

This structure, with 16 inspection centers, distributed in the north, south east and west regions of the city, has the capacity to carry out 700,000 inspections a month, on 211 inspection lines. Each of the centers has a minimum of 6 and a maximum of 32 lines, depending on the demand in each region of the city. Each line can inspect one vehicle at a time.

The program was established with the obligation for all vehicles to be inspected annually from the second year of their licensing. In other words, if a vehicle produced in 2009 was sold in 2010, it would have to undergo the mandatory annual inspection from 2011 onwards.

The vehicles were inspected in three stages. Initially, a visual inspection was carried out to check for the presence of blue or black smoke spewing from the exhaust pipe, the proper fixing of hoses and connections and consequent leaks of oil, fuel, brake or radiator fluids, the absence of a catalytic converter on vehicles approved with this equipment (the information can be found on the vehicle's technical data sheet, available to the inspector when the vehicle is analyzed), noticeably irregular engine operation, among other observations.

The main purpose of this first stage of visual analysis (called pre-inspection) was to exclude vehicles in a visibly poor state of repair from the inspection, giving the owner the opportunity to properly maintain their vehicle and then return for the inspection itself without having the vehicle fail.

Once the vehicle had passed the visual pre-inspection stage, it went on to be measured for gas emissions (CO and HC for petrol, ethanol or flex-fuel vehicles, including motorcycles) and smoke opacity for diesel vehicles.

The noise level analysis was then carried out on a sample basis, chosen at random or on the recommendation of the technician who carried out the pollutant emissions inspection, in the event of a suspected noise level above normal during the accelerations required for the emissions test.

Noise analysis was carried out in accordance with IBAMA Normative Instruction 06/2009.

If the vehicle failed the first vehicle inspection, the owner had to send it in for proper maintenance, rectifying the problems pointed out in the inspection, and return for a new test. This new inspection was called a reinspection and had to be carried out as many times as necessary until the inadequacies were corrected and the vehicle was approved.

Vehicle models sold in Brazil must be certified by IBAMA, which certifies that their design and manufacture comply with legal environmental requirements.

In order for these emission parameters to be met throughout the vehicle's useful life, it is necessary to comply with the maintenance plan recommended by each manufacturer for each type of vehicle.

The main objective of the I/M-SP was to check whether the vehicles in circulation in the municipality respected these emission standards during their useful life.

All manufacturers recommend that vehicles be serviced every 10,000 km or every year of use, whichever comes first (AFEEVAS).

Considering, by analyzing the statistical data produced by IM-SP, that the average annual circulation of each vehicle (gasoline and ethanol vehicles) is 15.000 km in the city of São Paulo (CONTROLAR - Annual Reports 2008 to 2012), every vehicle should undergo at least one overhaul every year, which justifies the I/M programs requiring also an annual inspection, in order to gauge the quality of the overhauls carried out throughout its useful life and effective compliance with the legal requirements regarding the maximum volume of pollutant emissions. For diesel-powered vehicles, the average annual mileage is over 20,000 km, justifying even more mandatory annual inspections. It is important to note, as mentioned at the beginning of this chapter, that the requirement for vehicles to be checked annually in I/M programs is included in federal legislation.

This need for annual checks also becomes clear when we analyze the emissions data of vehicles inspected and failed by the I/M-SP: the degradation of the environmental quality of these vehicles without maintenance or with inadequate maintenance, that is, the increase in the volume of emissions tends to occur from one year to the next, due to the natural degradation of the fleet, making it evident that, after the re-inspections provided for in the program, the fleet returns to a total emission volume similar to the final result of the previous year.

This shows us that some vehicles are taken to the annual inspection without prior maintenance and, if they fail, they undergo maintenance and then pass the next re-inspection. This phenomenon occurs every year, demonstrating the importance of checking the entire fleet annually.

Within the scope of the São Paulo program, diesel vehicles were initially analyzed (2008) and the following year the cumulative inspection of gasoline and ethanol vehicles with up to 5 years of use began, which correspond to approximately half of the circulating fleet.

The municipality's choice to start the program with this newer fleet was due to the fact that these are the vehicles that are most susceptible to lack of maintenance, as they are manufactured with devices (such as a catalytic converter) that retain up to 95% of the pollutants emitted by the engine. In other words, a vehicle with this technology that has not been properly maintained or, worse, that has been tampered with by removing the catalytic converter, for example, can emit 20 times more pollutants than it could if it had been properly maintained (CONTROLAR Annual Report 2008). What's more, these vehicles are easier to bring back into line with the original specifications, as diagnostics are carried out using electronic equipment that is widely available in the workshop network, as are spare parts.

Since 2010, the inspection of the entire fleet registered in the city has been required, broadening the scope of inspections and including the analysis of older vehicles.

The purpose of the inspection and maintenance programs is to monitor the volume of pollutants emitted by vehicles in circulation, demanding that, in order for a given vehicle to pass the inspection test and consequently be authorized to circulate, its conformity with the original specifications certified by IBAMA, in order to prevent users from compromising the conformity of their vehicles with those required by environmental legislation due to lack of maintenance or inadequate maintenance.

In practice, the implementation of the I/M-SP by the São Paulo PCPV led to car and motorcycle owners having to maintain their cars and motorcycles under penalty of not being authorized to circulate if they failed the annual test or if they failed to show up at the inspection centers to take part in the tests.

This is the great benefit of this type of program: to promote and consolidate the culture and habit of preventive and corrective maintenance in the population that owns motor vehicles and in workshops, in order to control the avoidable increase in emissions of atmospheric pollutant gases and noise from the circulating fleet.

In addition, permanent vehicle fleet inspection programs have the benefit of allowing the government to gain a deeper understanding of the fleet, including identifying manufacturing or maintenance problems inherent to certain makes and models, which can generate a database offered to the industry or repair network, as the case may be.

In this context, periodic inspections, such as the IM/SP program, have proven to be the best way of controlling emissions from fleets of motor vehicles in circulation, and this mechanism has been adopted by more than 50 countries.

The following tables, taken from the presentation *"Environmental Benefits and Results of the I/M-SP Program*" made by the author of this paper at the SIMEA 2013 congress in São Paulo, based on statistical data produced by the Municipal Transport Secretariat (SMT), give us the exact dimension of the benefits produced by the I/M-SP program during 2012.

The I/M program in the city of São Paulo established that vehicles that met all its requirements would be considered approved and entitled to a license to circulate.

Vehicles that did not meet the requirements could be separated into two groups: Rejected vehicles were those in which the test was carried out and identified that one or more pollutant emission parameters had been exceeded; Rejected vehicles were those in which the pollutant emission test was not carried out, because some abnormality was identified in the vehicle before the test was carried out that made it ineligible for approval (for example, it was missing the cover of the engine oil tank or the fuel compartment).

Once the emissions test had been carried out, if the vehicle did not pass, the owner had to have it serviced and returned for a new test, the re-inspection.

In the figure, the final re-inspection is the last inspection carried out on the vehicle in a given financial year, the result of which can be a rejection, failure or approval.

Figures 8, 9 and 10 were assembled from data extracted from the CONTROLAR annual reports.

Figures 8 and 9 show the percentage of vehicles that passed, failed and were rejected in the first inspection and the percentage of vehicles that passed, failed and were rejected in the final inspection, demonstrating the evolution in quality of the fleet at the beginning and end of the inspection period. The years indicated in the graphs refer to the years in which the vehicles were manufactured.

Figure 8. Initial inspection results, 2012

Source: CONTROLAR, 2012

Figure 9. Results of the 2012 final inspection

Source: CONTROLAR, 2012

Figures 10, 11 and 12 show the average emissions of the most relevant pollutant emitted by each category of vehicle (for cars, CO emissions; for motorcycles, HC emissions and for diesel vehicles, particulate matter emissions) of those vehicles that passed their first inspection; the average emissions of the vehicles that failed their first inspection and the average emissions of the vehicles that failed their last re-inspection, i.e. those that failed or were initially rejected but had one or more re-inspections, regardless of whether the final result was a pass or a fail. In each category, the data is grouped according to the technological stage of the vehicles used by PROCONVE, as shown in the table below each figure.

Figure 10. CO averages - gasoline, ethanol and flex-fuel cars

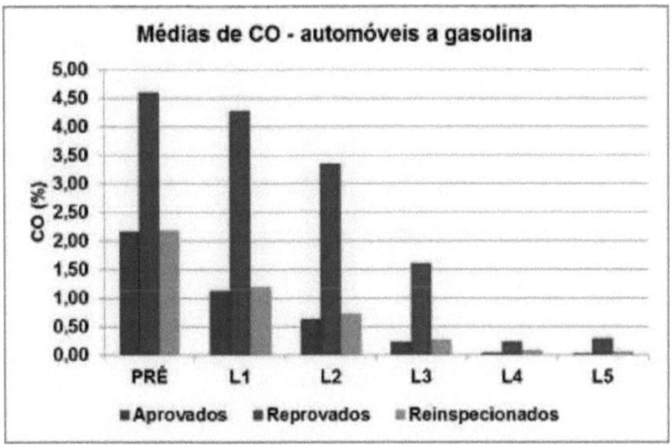

Source: Environmentality 2012

Figura 10 Petrol, ethanol or flex-fuel cars and vans:

- Pre-Proconve - Vehicles produced up to and including 1988
- L1 - Vehicles produced from 1989 to 1991
- L2 - Vehicles produced from 1992 to 1996
- L3 - Vehicles produced from 1997 to 2005
- L4 - Vehicles produced from 2006 to 2008
- L5 - Vehicles produced from 2009 onwards

Figure 11. HC averages - Motorcycles.

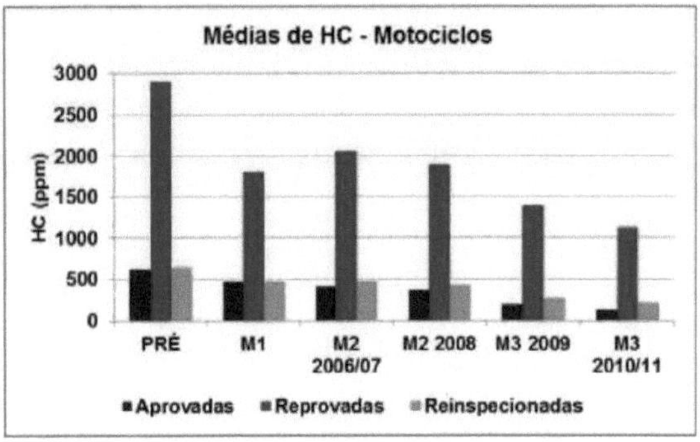

Source: Environmentality 2012

Figura 11 - Motorcycles (mopeds up to and including 50cc, motorcycles of any cylinder capacity, tricycles and quadricycles of any cylinder capacity).

- Pre-Promot - Produced up to and including 2002
- M1 - Produced from 2003 to 2005
- M2 - Produced from 2006 to 2008
- M3 - Produced from 2009 onwards

Figura 12. Average emissions (MPmg/Nm^3) - diesel trucks

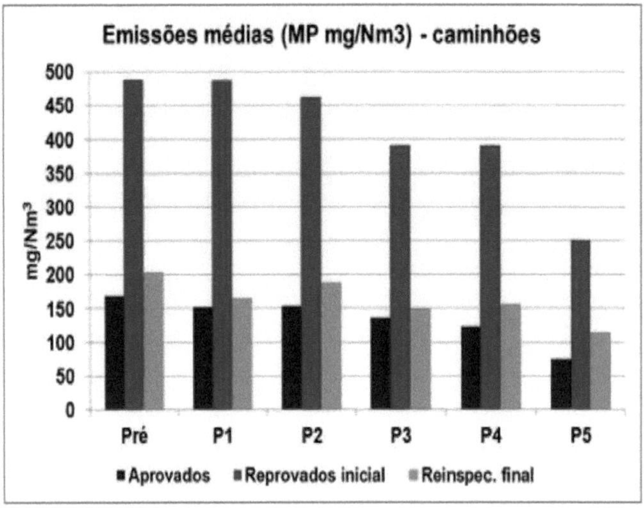

Source: Environmentality 2012

Figura 12. Diesel trucks

- Pre-Proconve - Produced up to and including 1988
- P1 - Produced from 1989 to 1993
- P2 - Produced from 1994 to 1995
- P3 - Produced from 1996 to 1999
- P4 - Produced from 2000 to 2005
- P5 - Produced from 2006 onwards

The total reduction in emissions (environmental gain) since the IM/SP Program began to date has been as follows (ENVIRONMENTALITY, 2012).

- CO emission reduction - Otto cycle engines - 49%
- Reduction in HC emissions - Otto Cycle - 39%
- Reduction in HC emissions - Motorcycles - 42%
- Reduction in annual CO emissions - Motorcycles - 34%
- Reduction in the annual emission of Particulate Matter - Diesel - 28%

The IM/SP experience also shows that there is an accumulated gain, added to the quality of the fleet, in other words, statistical analysis certifies that a degree of improvement in the average environmental quality of the fleet was incorporated after the program was implemented, even taking as a basis the measurements made only in the first inspection. This phenomenon is demonstrated in Figure 13, which shows that the average emissions at the start of the second and third years were lower than those found in the first year of the program's implementation, configuring a gain that can be considered permanent for as long as the Program existed (obviously the suspension of the Program can lead the population to abandon the habit of preventive maintenance again, returning to the condition prior to its implementation). It also shows the annual gain achieved by continuing with the program.

Figura 13. Average carbon monoxide emissions (%) - OTTO (Petrol, Ethanol and Flex)

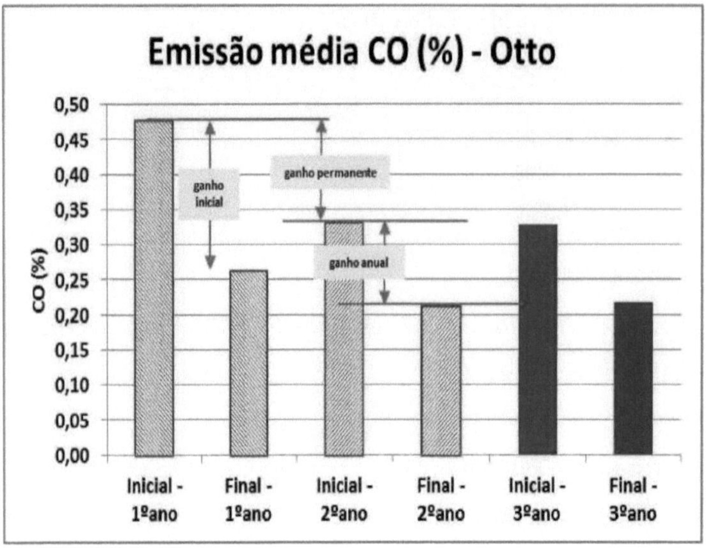

Source: Environmentality 2012

From the statistical data produced by I/M-SP, it is possible to demonstrate the improvement in the environmental quality of the city's vehicle fleet, as the emission volumes of each vehicle that took part in the program are calculated. The data is very consistent, as it was obtained from the study of more than 15 million inspections already carried out by the program, consolidated in the annual reports of the Concessionaire (CONTROLAR).

6. AUTOMOTIVE TECHNOLOGY AND FUEL QUALITY

Fuel burns by oxidizing the carbon in its molecules which, depending on the degree of oxidation, can turn into carbon dioxide, carbon monoxide, free carbon or other organic compounds. The composition of fuels includes elements other than carbon and water, such as nitrogen and sulphur, which are released into the atmosphere in greater or lesser proportions at the time of combustion, giving rise to atmospheric pollutants. Nitrogen is sometimes not present in the composition of the fuel, but because it is an element in greater proportion in the air, it combines with oxygen during combustion, giving rise to various compounds (NOx) that did not exist in the fuel or in the atmosphere before combustion, but which are highly toxic (Branco, 2004).

Other components in the composition of additives, whose function is to improve combustion, can be toxic (such as tetraethyl lead, added to gasoline) (Branco, 2004).

Another issue strongly related to the type of fuel is the use of fossil fuels or renewable fuels such as ethanol.

The choice of using fuels such as ethanol has the advantage (from an environmental point of view) of capturing the carbon dioxide emitted when burning the fuel through the production of photosynthesis during the growth of the crop, in this case sugar cane.

In this sense, the big problem (with regard to carbon dioxide) lies in the burning of fossil fuels, which releases carbon molecules into the atmosphere without the corresponding consumption of these molecules by photosynthesis.

The quality of the fuels, with attention to the approval standards defined by the ANP, distributed to supply the fleet must also be considered for an integrated emissions control policy.

Naturally, fuel adulteration must be combated vigorously, as the burning of any substances other than those specified in the vehicle's homologation will certainly have consequences for the functioning and durability of the engine and the gas after-treatment systems, which will be reflected in the quality of its emissions.

In the case of adulteration, solvents, which are more commonly used due to their low cost, release substances that increase emissions and can degrade catalysts, reducing or even nullifying their performance.

But the issue goes further than that: some stages of PROCONVE have required and will require the automobile industry, in order to meet its emission reduction targets, to adopt technologies that are only feasible with the supply of fuels of better environmental quality, with the reduction, for example, of the volume of sulfur and lead in fuels, without which it would not be possible to use catalytic converters, which would act innocuously or even deteriorate.

In addition, with the evolution of the environmental quality of engines, emission rates of some pollutants that were apparently acceptable in the past become extremely relevant over the years: if we take an engine produced and approved in 1994 as an example, a sulphur content of 2000 ppm (two thousand parts per million, the concentration found in metropolitan diesel oil supplied until 2004) increases particle emissions by 23%. For a 2005 engine, whose emissions are much lower, the use of this same diesel oil represents an increase in emissions of 80%, further justifying the use of fuel with a maximum of 500 ppm of sulphur (Environmentality, 2012).

For engines manufactured from 2009 onwards, the use of diesel oil with less than 50 ppm of sulphur is mandatory in order to maintain the approved pollutant levels and prevent the filters from clogging up, contaminating and compromising the functioning of the catalytic converter.

In other words, the technological development of vehicles must go hand in hand with the development of fuel quality. Without this compatibility, advances in the environmental quality of vehicles are of no use or cannot even be incorporated, at the risk of malfunction or premature wear.

Over time, the technology used in individual transport vehicles has advanced significantly, when analyzed from the point of view of comfort, performance, safety in use, reliability and pollutant emissions per liter of fuel used.

However, we can look at this development from another angle, which is the modification of the traction system of these vehicles, and from this point of view, little progress can be seen in practice.

The electric motor, a much more energy-efficient technology and immune to the loss of efficiency caused by traffic jams, is beginning to be seen as a possibility to replace traditional combustion engines only in the current era.

Focusing the analysis on Brazil or the city of São Paulo, it seems to us that this technology will still take a few years to become commercially implemented, since there are currently no 100% electric vehicles marketed on a large scale in the country, nor are there any incentives for their development or import.

In fact, the use of electric traction in vehicles has been known for over a century. As early as 1900, there were more than 1,500 electric vehicles on the road in the United States, competing with gasoline and steam vehicles. The number was quite significant in a market where just over 4,000 vehicles were sold that year throughout the country. At the time, the market for these electric vehicles seemed very promising, especially because of their smooth traction and, according to Santos Dumont, their reliability. The then automobile entrepreneur William C. Whitney bought a taxi company on the verge of bankruptcy with the aim of transforming it into a large electric taxi service company, renamed the Electric Vehicle Company, with its market initially focused on New York City (Branco, 2009).

The main problem pointed out today for the widespread use of electric vehicles, autonomy, was also the main issue to be solved back then, with all due respect: electric vehicles had to be refueled every two hours. The solution found was to replace the discharged batteries with fully charged ones for each autonomy period, instead of charging them. In this way, the vehicles could be driven all day long, with short breaks to change the batteries, an operation that was naturally much quicker than recharging the batteries, which required several hours when the vehicle was stationary.

The failure of the plan to implement these electric taxis was precisely due to the difficulty of maintaining these batteries, which had a wide range of defects, requiring a much larger stock of units than their creators had initially anticipated and, consequently, much higher costs. Before long, the Electric Vehicle Company was operating a fleet that didn't meet its operating requirements, vehicles that broke down a lot, and the company opted to renew its fleet with gasoline-powered vehicles. This episode is attributed to people's resistance to the use of electric vehicles, which led to the disuse of this technology at the time.

The brief account of this episode in the car industry shows us how much a certain technology can be replaced for reasons that are not related to its quality. Market issues are often more important than a well-structured technical analysis of a particular technology or product.

The automobile market doesn't seem to be oblivious to these subjective evaluations made by consumers or users. On the contrary: research carried out in the United States by the car manufacturer Toyota (interview with Gabriel Murgel Branco, consultant at Toyota do Brasil, 2013) showed that one of the items taken into account when consumers choose certain categories of vehicles is a strong engine noise, which is valued by a certain segment of consumers. This characteristic, non-existent in an electric car, led the company to consider the need to create an "artificial" noise for its electric vehicles, imitating that emitted by large combustion engines.

Looking back, it seems clear to us that issues relating to the emission of pollutants by vehicles cannot simply be left to the analysis of the industry and user taste, and it is essential that emission standards are increasingly restrictive, in order to force the automobile industry to seek technological solutions that lead to these new (and lower) emission standards.

Certainly, the evolution of vehicle environmental programs, with ever lower emission standards, will force the use of hybrid technologies, with greater or lesser support from electric motors and systems for using the energy wasted during braking. In other words, regardless of market issues, the technologies that are most successful in reducing emissions should certainly be used by the automotive industry. This reinforces the assertion that industry control programs are absolutely indispensable.

7. CONCLUSIONS

The transport sector is currently responsible for more than 75% of the world's air pollution (WHO).

At the end of the 1980s, Brazil implemented PROCONVE, which regulated the production of the automobile industry with regard to the emissions of polluting gases generated by vehicles. The program developed in several phases, gradually becoming more restrictive, so that a vehicle produced today emits, for example, 20 times less CO than a vehicle manufactured in the early 90s.

But this great environmental program needs constant adjustments to cope with the technological advances perceived by the automobile industry, because if PROCONVE was the legal regulatory framework that forced the automobile industry to worry about the emission of pollutants from its products, technological evolution was the way to meet this legal requirement.

As an example, when PROCONVE was conceived, there were no limits on pollutant emissions for hybrid electric vehicles, a technology that did not exist commercially at the time. Naturally, not even the way in which vehicles using this technology would be assessed was considered, which is currently a question that raises a number of doubts: under what conditions should the hybrid vehicle be tested? With the electric motor in action or considering only the production of gases from the combustion engine? What is the most suitable cycle for measuring emissions? And as soon as each of these questions is answered, many others will certainly arise, due to the constant technological evolution of vehicles. The hybrid vehicle itself already has more than one technology: parallel engines, series engines, plug-in engines, etc.

These questions, if not adequately studied and answered in order to improve the legislation governing PROCONVE, may on the one hand be a source of release for new pollutant emissions, but they run the serious risk of discouraging technological progress, insofar as the legislation is not capable of incorporating a technological change with all the benefits involved.

At the same time, it is necessary to control the emission of polluting gases from vehicles circulating in cities, obliging the volume of polluting gas emissions to remain within previously established parameters, consistent with those perceived at the time the vehicle was produced and put into circulation.

In this respect, it is essential to implement Inspection and Maintenance programs - I/M, similar to the one implemented in the city of São Paulo, which has led to substantial gains in the quality of the city's vehicle fleet.

Sâo Paulo's I/M Program has proven to be highly efficient, responsible for a significant reduction in the emission of pollutants from mobile sources and reducing the number of days of inadequate air quality not only in the city of Sâo Paulo, but also in the Metropolitan region.

In addition to the great environmental benefit provided by the Program, the volume of data generated allows for the constant improvement of environmental policies and communication with the population, making it an efficient tool for social communication.

While these aspects directly linked to the environmental quality of vehicles are extremely important in contributing to air quality in cities, the greater concentration of journeys made on public transport systems and the corresponding reduction in the volume of individual transport (modal rebalancing) can be the source of enormous gains in atmospheric quality, insofar as it promotes a huge reduction in the volume of atmospheric pollutant gases emitted per passenger transported.

Energy expenditure allows for similar reasoning: if technological advances in the design of manufactured vehicles can represent a huge gain in energy use, moving towards hybrid electric and pure electric technologies, the use of these technologies in public transport can boost this reduction in energy use, as they are capable of transporting many more passengers per unit of energy used.

Therefore, an adequate policy for reducing emissions in the transport sector cannot do without one of these three fundamental pillars: the production of vehicles with environmental quality, with the lowest emissions of pollutants that can be achieved by existing technology and that are economically viable; environmental control of the fleet in circulation, through periodic vehicle inspections and, finally, the constant search for modal balance, offering public transport as an alternative to individual transport, allowing the level of 70% of people's motorized transport being carried out by public transport systems to be reached.

BIBLIOGRAPHICAL REFERENCES

The European transport policy for 2010. [Technical Document]. European Commission. Office for Official Publications of the European Communities. Luxembourg, 2001.

AFEEVAS. *Vehicle emission limits and fuel specifications* - South America. Sao Paulo, 2012.

ANALYSIS OF URBAN MOBILITY: SPACE, ENVIRONMENT AND EQUITY. [Technical Document]. Development Bank of Latin America, 2011.

ANP. *Brazilian Statistical Yearbook of Petroleum, Natural Gas and Biofuels 2012.* Brasilia: ANP, 2012.

ANTP. *Report* suggestions presented at the 18th ANTP congress. Rio de Janeiro, 2011

BRANCO, A.M.; BRANCO, G.M. *A alternativa dos ônibus a hidrogênio na RMSP*. Sao Paulo: UNDP/GEF, 1998.

BRANCO, A.M. *History of public transportation in São Paulo.* Sao Paulo: AM Branco Consultores Associados, 2008.

BRANCO, A.M. *Trams, trolleybuses and the evolution of electric transportation.* Sao Paulo: AM Branco Consultores Associados, 2009.

BRANCO, G.M.; BRANCO F.C. *Inventory of mobile sources*: prospective and retrospective analysis of the benefits of PROCONVE for air quality from 1980 to 2030. Study for the Ministry of the Environment and the Hewlett Foundation. Sao Paulo, March 2007.

BRANCO, G.M.; SZWARC, A.; BRANCO F.C. *Programa ambiental de inspeçao e manutençãoçâão veicular:* principios, fundamentos e procedimentos de teste. Sâo Paulo: Edgard Blucher, 2012.

BRANCO, G.M.; BRANCO, F.C.; VASCONCELLOS, E.A.; FERREIRA, A.L. Environmental impacts of mobility evolution in Brazil. In: *Symposium on Future Challenges of Transport and Environment.* Berlin, June 2008.

BRANCO, M.C. *Evaluation of the Toyota Prius vehicle and its feasibility for use in the taxi fleet in the city of São Paulo.* SMT Report. Sâo Paulo, Dec. 2011.

BRANCO, M.C et al. Environmental benefits and results of the I/M-SP program In: *XXI SIMEA*, 2013. Sâo Paulo, 2013

BRANCO, S.M. *Ecossistêmica:* uma abordagem integrada dos problemas do meio ambiente. 2. ed. Sâo Paulo: Edgard Blucher, 1999.

BRANCO, S.M. *Energia e meio ambiente*. Sâo Paulo: Editora Moderna, 2004.

BRANCO, S.M.; MURGEL, E. Poluiçâo *do ar*. 2 ed. Sâo Paulo: Editora Moderna, 2009.

CERTU. *Nouveaux Systémes de Transports Guidés Urbains de CERTU*. Centre d'Etudes sur les Réseaux, les Transports, L'urbanisme et les Constructions Publiques du Ministére de L'Équipement des Transports et du Logement: France, 2000.

CET. *Survey* prepared for the evaluation of the Pedestrian Protection Program: Sâo Paulo, 2011

CINTRA, M. *Os riscos de Sâo Paulo:* uma agenda sobre mobilidade, Sustentabilidade convivência e gestâo. Sâo Paulo: CPV Editora, 2008

CITIGROUP. Study "Off the beaten path". [s.l], 2013.

COLETIVO. *Technical Magazine of Sâo Paulo Transportes*. SPTrans. Issue Zero. Oct. 2011.

COLETIVO. *Technical Magazine of Sâo Paulo Transportes*. SPTrans. N.1. Dec. 2012.

CONTROL. IM/SP annual monitoring *reports* - 2008, 2009, 2010, 2011 and 2012.

COMPET. *Discussion of consumption and energy efficiency indices for labeling Brazilian vehicles*. Rio de Janeiro: COMPET, 2006.

DALLARI, A.A.; BRANCO, A.M. *The financing of public works and services*.

Sâo Paulo: Paz e Terra, 2006.

DIRECTIVE 2009/33/EC of the European Parliament and of the Council. *Official Journal of the European Union,* 2009.

ENVIRONMENTALITY. *Report* analyzing the effects of PROCONVE 1992/2012. São Paulo, 2013.

GLAESER, E.L.; *Urban centers: humanity's greatest invention.* Rio de Janeiro: Elsevier, 2011.

KIM, W. *International program on urban environment policy for sister city and friendship city officials*. Seoul Human Resource Development Center. Seoul, 2010.

METRO/SP - Companhia do Metropolitano de Sao Paulo. Origin and destination *survey* 2007. Sao Paulo, 2007.

McKinsey & Company. Study *report* on Brazil's air transportation sector. Rio de Janeiro: McKinsey & Company, 2010.

WHO. *Report.* World Health Organization. International Agency for Research on Cancer. Washington: WHO, Oct. 2013.

OECD. *Transport and environment.* Paris, 2008.

PREFEITURA DE SÂO PAULO. *Building sustainable cities:* summary report of the C40 Sao Paulo Climate Summit 2011. Sao Paulo, 2011.

SAE. SAE Automotive Engineering *REPORT.* Sao Paulo, Sept. 2004.

SALDIVA, P.H.N. *Poluiçâo atmosfèrica e saù*, uma abordagem experimental. Sao Paulo: Greenpeace, 1998.

SALDIVA, P.H.N. et al. *Meio ambiente e saùde*: o desafio das metrópoles. Sao Paulo: Ex Libris, 2010

SALDIVA, P.H.N. et al. *Vehicle Emissions Control Program.* PROCONVE. Emissions of atmospheric pollutants by mobile sources and estimation of the effects on health in the RMSP: current scenario and projections. laboratory of experimental atmospheric pollution of the medical school of USP. Sao Paulo, 2007.

SECRETARIAT OF TRANSPORTATION OF THE MUNICIPALITY OF SÂO PAULO/SMT. Report: Retrofit Possibilities. Sao Paulo, 2011.

SOMMER, H. *Economic evaluation in the project* "Health Costs due to Road Trafic Related Air Pollution". Stockolm, jun. 2003.

SOUSA, E.L.L.; Macedo, I.C. et al. *Ethanol and bioelectricity:* sugarcane in the future of the energy matrix. Sao Paulo: Editora Luc, 2010

STIGLITZ, J. *Globalizaçâo:* como dar certo. Sao Paulo: Companhia das Letras, 2007.

TRANSPORT and the environment in Brazil. Publication of ANTP's Transport and Environment Commission. Sao Paulo: ANTP, Oct. 2011.

EUROPEAN UNION. *White Paper:* transport policy for 2010. Time to choose. Luxembourg, 2001

VASCONCELLOS, E.A. *Transporte e meio ambiente.* Sao Paulo: Annablume Publishing House, 2008.

WALSH, M.P.; BRANCO, G.M. et al. *Control of diesel vehicle pollution*: a strategy for progress in brazil . Study prepared for the Hewlett Foundation. Sao Paulo, 2005.

More Books!
yes
I want morebooks!

MIX
Papier aus verantwortungsvollen Quellen
Paper from responsible sources
FSC® C105338

Printed by Books on Demand GmbH, Norderstedt / Germany